Weather

by Emil Edwards

Leaves on these trees change color in fall.

Contents

Introduction: Weather All Around Us 4

Chapter 1

Big Idea Question

How Does the Sun Affect Earth? 6
- The Sun's Energy 8
- The Sun and Water 12
- Wet Weather 14

Chapter 2

Big Idea Question

How Does Weather Change? 16
- Day-to-Day Changes 18
- Changes in One Day 20
- Day Changes to Night 22
- The Four Seasons 24

Chapter 3

Big Idea Question

How is Weather Measured? 28
- Predicting the Weather 30
- Using Weather Tools 32
- Severe Weather 36
- Be Prepared 38

Conclusion: Weather on Earth 40

Glossary 42
Index 44

Introduction

Weather All Around Us

Sunny day

Next Generation Sunshine State Standards
SC.2.E.7.1 Compare and describe changing patterns in nature that repeat themselves, such as weather conditions including temperature and precipitation, day to day and season to season.
SC.2.E.7.4 Investigate that air is all around us and that moving air is wind.

Weather is what the air is like outside. Sometimes the air is calm, and the sun shines. At other times, the air is breezy, and the sky is cloudy. Rain can fall.

You can use your senses to observe weather. You can see lightning and hear thunder. You can feel air move all around you. Moving air is wind.

Windy day

Rainy day

Chapter 1

Big Idea Question

How Does the Sun Affect Earth?

Next Generation Sunshine State Standards

SC.2.E.7.2 Investigate by observing and measuring, that the Sun's energy directly and indirectly warms the water, land, and air.

SC.2.P.10.1 Discuss that people use electricity or other forms of energy to cook their food, cool or warm their homes, and power their cars.

SC.2.E.7.1 Compare and describe changing patterns in nature that repeat themselves, such as weather conditions including temperature and precipitation, day to day and season to season.

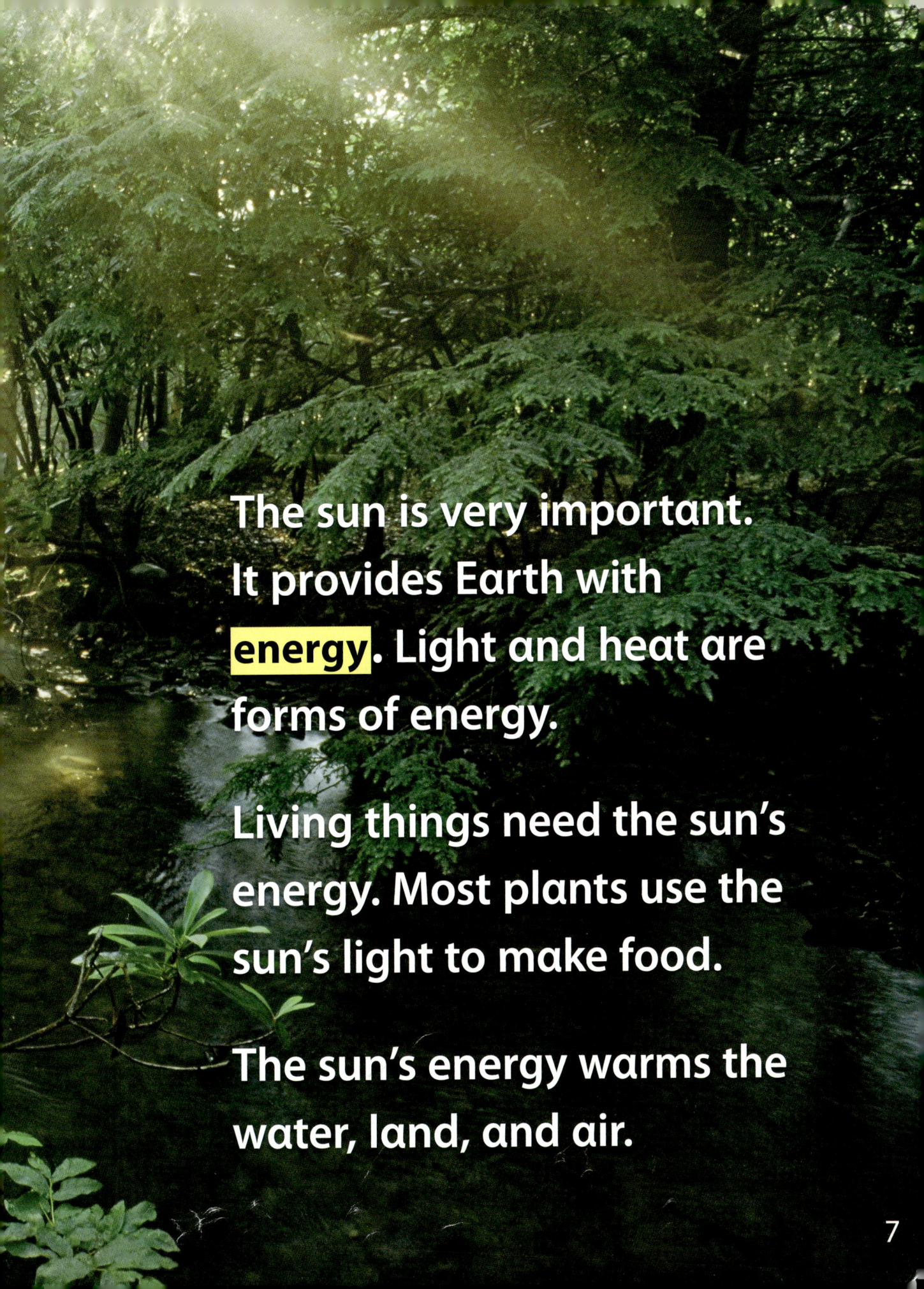

The sun is very important. It provides Earth with **energy**. Light and heat are forms of energy.

Living things need the sun's energy. Most plants use the sun's light to make food.

The sun's energy warms the water, land, and air.

The Sun's Energy

At noon, the sun is high in the sky.
More of the sun's energy warms
Earth at this time.

The sun shines on these trees from above.
The trees get a lot of energy from the sun.

In the morning and evening, the sun is low in the sky. Less of the sun's energy warms Earth at these times.

The **temperature** in a shady spot usually feels cooler than it does in a sunny spot. Temperature is how hot or cold something is.

The sun shines on these trees from the side.

The sun's energy is very strong. Some people use the sun's energy to heat or cool their homes and cook their food. The sun's energy can power cars, too!

This car uses the sun's energy to run.

This oven uses the sun's energy to cook food.

The panels on this house are called solar panels. They collect sunlight. They turn sunlight into a different form of energy. The energy is then used inside the house.

solar panel

The Sun and Water

Heat from the sun **evaporates** water into water vapor. Water vapor is a gas that rises into the air.

Water evaporates into the air.

The water vapor condenses and forms a cloud.

Then the water vapor cools and **condenses**, or forms tiny drops of water. The drops form a cloud. Water falls from the cloud. The cycle begins again.

Rain falls from the cloud.

Raindrops form puddles.

Wet Weather

Earth is surrounded by layers of air made from gases. Clouds form in the layer closest to Earth. Some types of clouds can bring wet weather.

Rain

Rain can fall when the air is hot, warm, or cool.

Freezing rain

Freezing rain is rain that freezes when it lands on a cold object.

Rain, freezing rain, snow, sleet, and hail are kinds of wet weather.

Snow

Snow can fall when the air is very cold.

Sleet

Sleet is ice mixed with rain and snow. It can fall when the air is cold.

Hail

Hail is balls of ice. It can fall when the air is warm or cold.

Chapter 2

Big Idea Question

How Does Weather Change?

Next Generation Sunshine State Standards

SC.2.E.7.1 Compare and describe changing patterns in nature that repeat themselves, such as weather conditions including temperature and precipitation, day to day and season to season.

SC.2.P.8.4 Observe and describe water in its solid, liquid, and gaseous states.

What is the weather like today? It might be cold in the morning and hot in the afternoon. It might be sunny or rainy. If it's sunny and rainy, you might even see a rainbow!

Day-to-Day Changes

Weather can change from day to day. Wind, temperature, and cloud cover can change. On cloudy days, there are many clouds in the sky. On partly cloudy days, the sun is partly hidden behind clouds.

Rain falls over the Golden Gate Bridge on a cloudy day.

You can observe how the weather might change. For example, Monday might be foggy and cold. Tuesday might be sunny and warm. On Wednesday, clouds might hide part of the sun. It might feel warm on a partly cloudy day.

Weather Changes from Day to Day

Monday	Tuesday	Wednesday
foggy	sunny	partly cloudy

Changes in One Day

Weather can change throughout the day, too. On a hot day, rain might cool the air. On a cold day, the sky might be clear in the morning. But heavy snow might fall by nighttime.

You can observe how the weather changes. The sky might change from light to dark before a storm. The air might change from calm to windy, too.

Calm

Breezy

Windy

Day Changes to Night

You can see the sun in the sky on most days. You cannot see the sun at night. This makes the sky light during the day and dark at night. Light days and dark nights is a **pattern**. It repeats over and over.

Day in Egypt

The sun's energy warms the air during the day. The air cools at night. This pattern of warmer days and cooler nights repeats, too.

Night in Egypt

The Four Seasons

Weather changes from season to season. Many places on Earth have four seasons. They are spring, summer, fall, and winter. The seasons follow a pattern. They come in the same order, year after year.

Spring

In spring, the days start to get longer. The temperature gets warmer.

Summer

Summer has the most hours of daylight. It is the warmest season.

As the seasons change, the temperature can change. The amount of rain or snow changes. The amount of daylight and darkness changes, too.

Fall

In fall, the days start to get shorter. The temperature gets cooler.

Winter

Winter has the fewest hours of daylight. It is the coolest season.

What are the seasons like where you live? If it is cold in winter, water can freeze. You might go ice-skating or build a snowman.

If it is hot in summer, cold water is refreshing. You might go swimming. You might make ice cubes and drink a cold glass of lemonade.

Chapter 3

Big Idea Question

How is Weather Measured?

Tim Samaras, National Geographic Explorer, uses a probe to measure the weather inside a tornado.

Next Generation Sunshine State Standards
SC.2.N.1.6 Explain how scientists alone or in groups are always investigating new ways to solve problems.
SC.2.E.7.1 Compare and describe changing patterns in nature that repeat themselves, such as weather conditions including temperature and precipitation, day to day and season to season.
SC.2.E.7.4 Investigate that air is all around us and that moving air is wind.
SC.2.E.7.5 State the importance of preparing for severe weather, lightning, and other weather related events.

Sometimes weather changes very quickly. Scientists use clues from nature to predict the changing weather. They also use **weather tools** to measure the weather.

Predicting the Weather

Scientists observe clouds to predict weather. Different kinds of clouds bring different kinds of weather. Rain rarely falls from cirrus clouds. But a lot of rain can fall from cumulonimbus clouds.

Wispy cirrus clouds do not bring wet weather.

Dark cumulonimbus clouds can bring a lot of rain or stormy weather.

Scientists also use the air temperature to help predict the weather. For example, on a cold, cloudy day, scientists might predict snow. On a warm, cloudy day, they might predict rain.

Puffy cumulus clouds are usually a sign of good weather.

Flat stratus clouds can bring snow or rain.

Using Weather Tools

Scientists use weather tools to measure the weather. A weather satellite measures the clouds in the sky. It measures how thick and how high clouds are.

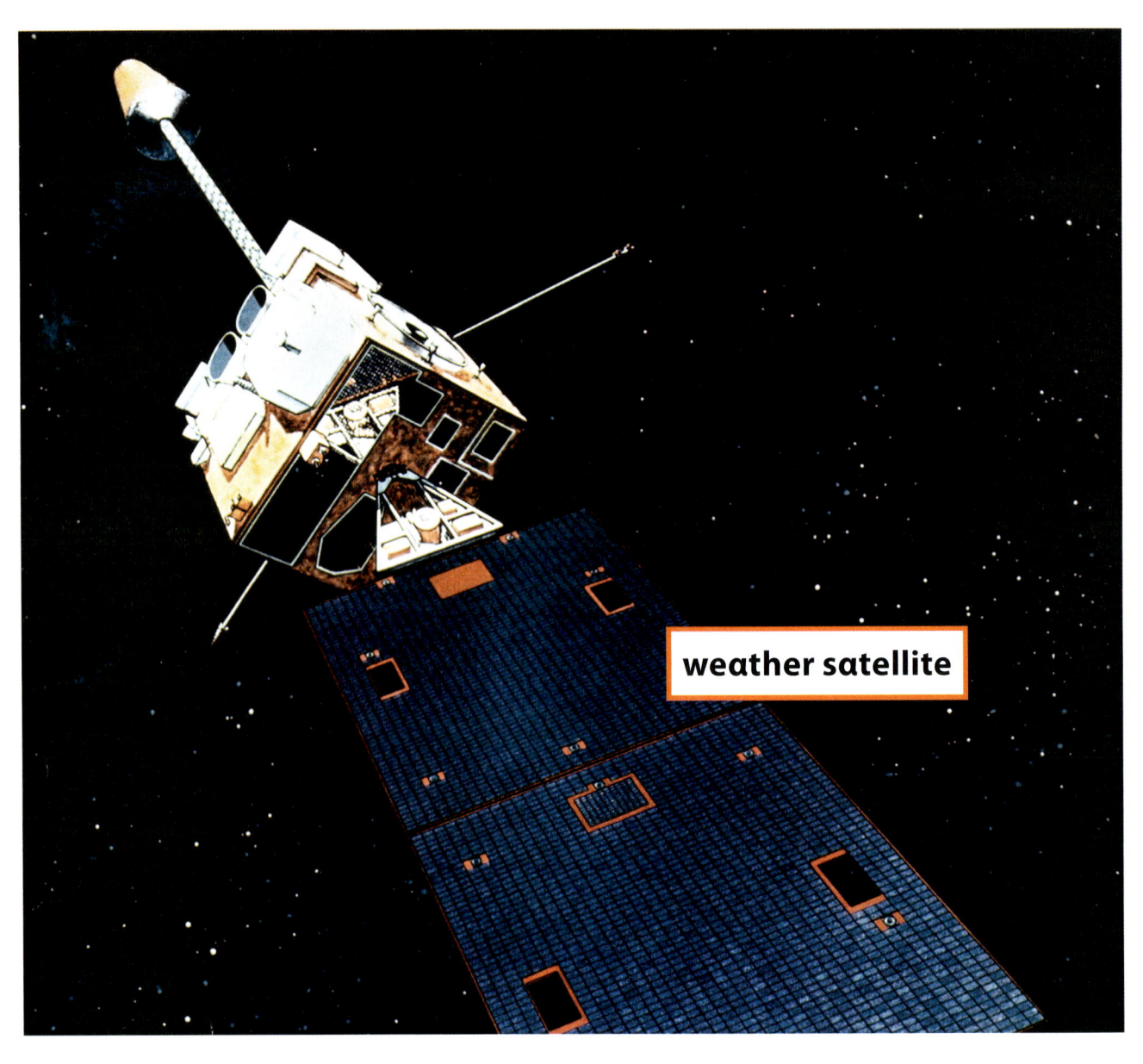

weather satellite

Some tools measure temperature or rain. A thermometer measures the temperature of the air. A rain gauge measures the amount of rain that falls.

thermometer

rain gauge

33

Some weather tools measure the wind. An anemometer measures wind speed, or how fast the wind is blowing.

anemometer

A wind vane and a windsock measure wind direction, or where the wind is coming from. The wind can blow from the north, south, east, or west.

wind vane

windsock

Severe Weather

Weather tools help measure severe weather. The sky turns gray and strong winds blow. Thunder booms and lightning streaks across the sky. Rain and hail fall from the clouds. This is severe weather.

Thunderstorm

Tornado

Severe weather can come in the form of lightning, strong winds, tornadoes, hurricanes, and blizzards.

Hurricane

Blizzard

Be Prepared

Severe weather can be dangerous for people and other living things. During severe weather, it is important to stay in a safe place indoors.

Go down to the basement when there is a tornado warning.

Leave your home quickly with your family when there is a hurricane. Follow posted signs.

Stay away from water and things that use electricity during a lightning storm.

Stay inside during a blizzard. Be sure to have plenty to eat and drink in your home.

It is important to be prepared, too.

- Put together an emergency kit with water, food, a flashlight, a radio, and medical supplies.

- Meet with your family to create a safety plan.

- Practice your plan often.

Conclusion

Weather on Earth

Living things need the sun to survive. The sun's energy warms Earth. It affects the weather.

The weather on Earth is always changing. It changes from day to day and from season to season.

Scientists observe clouds and use weather tools to predict how the weather might change. Scientists use tools to measure the weather, too.

Next Generation Sunshine State Standards
SC.2.E.7.1 Compare and describe changing patterns in nature that repeat themselves, such as weather conditions including temperature and precipitation, day to day and season to season.
SC.2.N.1.6 Explain how scientists alone or in groups are always investigating new ways to solve problems.

Glossary

condense (page 13)
When water vapor cools, it **condenses** and changes from a gas to a liquid.

When water vapor cools, it can **condense** and form a cloud.

energy (page 7)
Energy is something that can change things or do work.

Heat and light are kinds of **energy.**

evaporate (page 12)
When water **evaporates,** it changes from a liquid to a gas.

Heat from the sun causes water to **evaporate.**

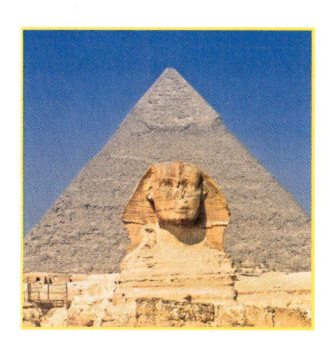

pattern (page 22)
A **pattern** is something that repeats over and over again.

Day turns into night. This is a **pattern.**

temperature (page 9)
Temperature is a measure of how hot or cold something is.

The **temperature** gets warm in the summer.

weather tool (page 29)
A **weather tool** is something scientists use to measure the changing weather.

This probe is a **weather tool.** It measures the weather inside a tornado.

Index

condense 13

energy 7–11, 23–25, 40

evaporate 12

fog 19

pattern 22–23

season 24–27, 40

solar panel 11

temperature 9, 18, 24–25, 31, 33

weather tool 29, 32–36, 40

Copyright © 2011 The Hampton-Brown Company, Inc., a wholly owned subsidiary of the National Geographic Society, publishing under the imprints National Geographic School Publishing and Hampton-Brown.

All rights reserved. No part of this book may be reproduced or transmitted in any form or by any means, electronic or mechanical, including photocopying, recording, or by an information storage and retrieval system, without permission in writing from the Publisher.

National Geographic and the Yellow Border are registered trademarks of the National Geographic Society.

National Geographic School Publishing
Hampton-Brown
www.NGSP.com

Printed in the USA.
RR Donnelley, Jefferson City, MO

ISBN: 978-0-7362-7555-2

11 12 13 14 15 16 17

10 9 8 7 6 5 4 3